Car Buying Guide

Save Time and Money By Learning How to Find the Best Car Buying Opportunity and Negotiate the Best Price While Avoiding the Car Dealer's Attempts to Get the Most Money Out of You

By Zack Keever

Contents

Thank you for buying this book and I hope that you will find it useful. If you will want to share your thoughts on this book, you can do so by leaving a review on the Amazon page, it helps me out a lot.

Chapter 1: Reasons for Purchasing a Car

Besides the apparent reason, that being transport, much like water, a car is now seen as a requirement in a lot of metropolitan locations. Among the reasons to think about when purchasing a car is if the car being bought is a bargain. Is it in good shape? How long does it have left? Is it an outstanding ride? Does it have a terrific warranty, and so on. As we stated, it's practically inconceivable to get along without having a car, and in case you can't pay for a brand-new car, here are certain standard reasons you must think about when making that necessary purchasing decision.

Is It a Bargain?

The ideal time to purchase a car is later on in the month. Typically, there are many bonuses and rebate programs provided by month's end based upon a dealership's month-to-month sales quotas. The idea is, in some cases, dealerships fall short of fulfilling their sales goals if their performance is evaluated by the month's end. This takes place more often than not. The outcome of this is excellent news for the majority of car purchasers. Because dealers want to hit their quotas a lot more desperately, they have no choice except selling the car at a lower than regular price simply to get the sales they require to fulfill their targets.

Peace of Mind

Often dealerships are going to do a "dog and pony show" with the goal of making a sale. The outcome is that car purchasers then wind up with features that they recognize later on they do not actually require with a price they really might not pay for. The point? Don't purchase more than you certainly require. Stay away from unneeded stress and/or sleep-deprived nights.

Always keep in mind to believe your instincts. If you are uncertain, the response to that is do not-- do not allow yourself be pushed into buying now. This is the typical technique of dealers where they would persuade the consumer to purchase that car, on the spot, this second, today!

Remember that you can always sleep on it and after that choose. This is so much better than deciding now and forever holding your peace. In case the vehicle is gone the following day, there's most likely a great reason why you should not have had it, to begin with.

First Impression

First impressions generally last. Although not always accurate, in purchasing cars, you can collect a good deal of information by simply seeing how car sales representatives deal with you. Typically, these non-verbal signals are sufficiently accurate to provide you an excellent idea of what to anticipate from the sales representative and your vehicle buying experience.

In case you sense that the individual offering you the car is not dependable, or you just do not like him/her, always keep in mind that you have the choice to go away. And, time is valuable, particularly yours. So do not squander it. You are constantly free to leave nicely, anywhere, anytime, and in any way you like it. However, if there are hardly any dealerships in your location, you might wish to stick around for while longer and wait until the negotiations you have with your dealership breaks down and you just have no alternative but to leave.

The Price on the Invoice

Another thing to think about when purchasing a car is the invoice price. Make certain that you ask for an invoice. In case dealers are working their tails off against it, you have ample reason to think that the vehicle in question is an undesirable deal.

Generally, the invoice price is the expense the dealer paid the producer for purchasing the vehicle. This is prior to any incentives or rebates being involved. When this is understood, just then are going to have an idea regarding just how much the dealer might profit from every car sold. Understanding this information might allow you to feel more confident in working out the price.

You might likewise stumble upon this periodically when buying a "secondhand car." This would happen mainly when you are purchasing a really late-model second-hand car still under the factory warranty and with low mileage.

The Producer's Suggested Retail Price

Simply put, this is typically referred to as the sticker price. This is the cost one typically finds on the window of the vehicle being offered.

Never ever think about paying the money specified on the sticker. This amount is simply the beginning point to your negotiations.

Incentives

There are examples when producers supply the dealerships with something extra, such as cash, bonus or rebates due to the fact that they had the ability to sell cars which are either undersold or overstocked. Ensure that prior to really buying a car, you had the ability to understand if that vehicle you are purchasing has some dealer incentives linked to it. If so, remove that number from the car's buying price and have yourself an excellent deal and, ideally, a great car too.

Keep in mind the majority of our conversation on invoice pricing have to do with really late-model cars. It was worth pointing out here in case you do stumble upon the chance to buy this kind of "secondhand car."

Chapter 2: Where to Get Fantastic Second-Hand Cars

A franchised brand-new car dealership is a dependable, if not the sole source for anybody who wishes to purchase a brand-new car. However, to those who prefer purchasing previously owned or used cars, there are lots of choices and sources one might select from to be in a position to make that car transaction the ideal one.

Here are a few of the readily, along with easily offered resources to think about when purchasing a used car.

The Second-Hand Car Superstore

Picture it as the used cars' Wal-Mart. There have actually been many used car superstores which have actually mushroomed over the previous five years. These superstores typically have a big supply of used vehicles, and could range anywhere from 400 to 500 cars. The vehicles discovered in these shops generally are late models.

The cars discovered in these shops originate from auctions which cater specifically to car dealers. The great thing about purchasing from these second-hand car superstores is that the warranty they supply is frequently comparable to the warranty protection offered by dealers who sell brand-new vehicles. Nevertheless, it is still ideal to compare.

New Car Dealer

The reasoning here is that because brand-new car purchasers generally trade in their old vehicles when they purchase a brand-new one, the collection which new car dealerships have is usually substantial. So used car purchasers have a broader range of cars to pick from. Likewise, brand-new car dealerships offer much greater reconditioning to the used cars traded to them. They are likewise a more dependable source of used cars due to the fact that their company is more established in comparison to other secondhand car dealerships. Realize, though, that it would often require a great deal of negotiation on the price of the car to be bought. They might likewise try to make you purchase more than what you would, in fact, require.

Second-Hand Car Dealers

There are typically numerous used car dealers in each state. In some cases, they are little operations which don't have more than fifteen cars in a lot.

Established secondhand vehicle dealers, could have as many as hundred vehicles on the lot. The fantastic feature of purchasing from used car dealerships is that the prices they provide are a lot lower than new car dealerships. Additionally, it is relatively simpler to negotiate with them. Not so excellent news though, these cars' quality is generally lower than the brand-new ones, obviously. There likewise might be a tinier selection of models and makes.

Private Owners

The great thing about purchasing from private owners is that the prices they provide are generally affordable compared to dealerships due to the fact that this remains in accordance with book values. You likewise get the chance to, in fact, talk to the owner of the car and witness on your own how the car was or was not, looked after. A prospective drawback of this circumstance, nevertheless, is that it could be a tad bothersome driving to the private owner's location, particularly if you are thinking about taking a look at 8 various cars. Generally, that would be 8 various locations and appointments.

Be careful though, private owners with a staple of second-hand cars to sell might, in fact, be dealers. Do not hesitate to ask if you might potentially see the registration and title. Begin getting suspicious if you see it is just a couple of days old.

Auctions

The previous decade has actually seen the advancement of public auctions for cars. Initially, auctions such as these were reserved mainly for certified car dealerships. Now, even people have the chance to bid strongly for second-hand vehicles. The quality of these cars set up for auction, in addition to the selections of cars, in fact, differs from one auction to the next. There are those auctions which focus on late-model cars while there are others that commit themselves to more affordable and less costly cars. The great thing about purchasing secondhand cars from auctions is that you are able to easily contrast prices and cars due to the fact that they are revealed side by side with one another.

Likewise, the prices which they offer are probably lower than the one offered by dealers. Nevertheless, there is not much possibility for you to completely examine the car being auctioned off. And given that you are purchasing from an auction, it is for that reason recognized that all sales are final. Any vehicle bought is immediately your. Also, the bidding craze might take hold of anybody and there is a fantastic likelihood that you might pay way more than what a car really costs.

We are going to dive deeper into the topic of vehicle auctions in the next chapter.

Chapter 3: Financing

Prior to buying your dream car, whether brand-new or second-hand, attempt to evaluate if the budget plan can truly afford it.

Questions such as:

How are you going to pay for it?

Who is going to help you to pay for it?

What is the price limit?

How long is it going to require to pay off the car?

These ought to be thought about even while planning to purchase the car. The trend these days is that individuals purchase the cars by cashing out the deposit, and the balance is going to be paid by installment. Others are simply fortunate sufficiently to have actually saved the correct amount of cash that they have the ability to cash out the overall car cost, that, by the way, hardly ever takes place any longer. Never ever set aside the chance of paying thousands of dollars when purchasing from a dealership or a particular car business, where, ultimately, they earn more with interest, which requires years to pay off.

When you have actually discovered a method to fund the car you are interested in, then it's time to begin looking around. There are credit unions and even neighborhood banks that want to lend the required amount to acquire the car with an Interest rate of just 1.9 percent. Nevertheless, this might end up being a catch, given that this is going to just be taking place in the initial year. Without previous notice, these interest rates could go up, and that is total trouble for those with simply a fixed income each year.

It is a huge plus if a purchaser belongs to a credit union. Being a member might save you the hassle of spending an entire day in a lender's office since the processing of the loan might just take simply a couple of minutes after submitting the required documents for the request. In a credit union, 15 to 20 minutes is all that is required to do the application. They might even lend as much as $20,000.00 inside only an hour after signing the documents.

It could save you a fair bit of cash by researching before making that loan. Car dealerships are truly digging out the majority of the pennies in a purchaser's pocket by providing interest which is often unreasonable. There are 2 things to think about when thinking about financing a brand-new car:

Initially, what's the price you want to spend? Usually, asking yourself the question: Just how much of the car price do you plan to purchase instead? Just do so if you understand you can pay for the vehicle of your choice. Think about routine month-to-month expenses. The month-to-month payment for the brand-new car must to not get in the way of paying the fixed expenses. If it does, by simply doing the mathematics, then simply think about an older car or one which might not be quite as "upscale." Simply ensure that all is inspected and taken a look at to prevent an inconvenience in the future.

Next, is it truly essential to change cars every 2 to 4 years? Think about cars provided on a lease, if so. Other dealerships and car businesses provide the car leasing for that amount of time, that you could return; however, no cash is going to be reimbursed. Nevertheless, there is an allocated amount of range or mileage which ought to be covered throughout those years of a lease, however, this is negotiable. Meanwhile, if a purchaser is not thinking about changing cars, it's ideal not to go with the cars for lease.

It is best to obtain a loan from a financial institution or a credit union rather than a neighborhood car salesperson; they are going to absolutely attempt to squeeze your bottom dollar. Acquire information from somebody you can rely on and is a professional about funding a car, for they are going to have the ability to offer you suggestions to benefit your own interests. Funding either a brand-new or a used vehicle is a great deal of sweat, however, the determination to get the ideal vehicle at the ideal price could be seen as a success.

This ought to be a win, win scenario for anybody. Besides, it's your cash which is at stake. Carry out your research and it could be a valuable decision on your part.

Chapter 4: Purchasing a Used Vehicle

Purchasing a car from a person or from a neighborhood car dealership is one laborious task if the goal is to buy a second car which is still in great condition. A purchaser would never ever trade his hard-earned cash with just a total squandered used vehicle. Whether brand-new or utilized, when purchasing a car, it ought to be treated with identical value and importance.

There are things to take into account prior to purchasing the car and these are:

1. Inform yourself as a customer

2. Make a list of your requirements. Prioritize between the requirements and the desires.

3. Identify the budget plan and the type of vehicle that would ideally fit.

4. Select by types and models. Narrowing them is going to be valuable.

5. Do the research offline and online by dealers.

6. Understand the worth of used car in the market.

7. Research the Car history and identification numbers.

8. Take a mechanic when examining the vehicle whether purchasing from a buddy or a dealership.

9. Never ever avoid asking questions.

10. If there's a suspicion about it not being great, do not hesitate to leave.

Bringing a mechanic with you is constantly an excellent approach. Ensure your mechanic checks the overall vehicle, from the history down to the final screw. It must be parked on a level spot. You ought to ensure that it was driven for around an hour prior to the assessment.

Inspecting the exterior.

Walk around to see if the body is damaged in any way. The corners of the vehicle ought to be shaken and bounced back and forth to observe if the shock absorbers still work well. Ensure that the wheel bearings do not generate any noise when attempting to yank the front tires by pulling them. Open the doors, raise the trunk and the hood to uncover if all the rubber seals are still there. This is going to likewise inform if there's something loose around the hinges. Look for indications of a repaint. There is going to be a distinction in the color due to the fact that stores can never ever replicate the initial paint of the vehicle. Ask somebody to switch on the lights outside the vehicle and see if all are functioning properly. Cars with just 30,000 miles of travel need to still have its initial tires. Be careful if you discover a car with just a couple of miles of travel but with brand-new tires. When the test drive is finished, examine the discs of the brakes, this ought to still be tidy and smooth. Examine the windshield for fractures.

Inspecting the Interior.

Even though it seems odd, smell the car interior. Smell beneath the carpet and mats. If you smell mildew, then it's an indication that there is a leakage someplace or that the vehicle might have been damaged by the flood. Turn on the air-conditioning to make certain that it actually turns the entire interior of the car cold. After this, attempt the heater. Experiment with all lights within too, and always remember to blow the horn. Likewise, try all the seat changes. The upholstery ought to still remain in good condition as well. There is a great deal more to inspect and here the mechanic could assist the purchaser.

Looking Within the Trunk

Again, smell the insides and look for any indications of leakages. Make certain that some fundamental car tools are still there for the new user.

Inspecting Under the Hood

Feel the circuitry for any brittleness or cracks. Squeeze the fan belt and hoses for any cuts and feasible electrical tape patches. Do not remove the radiator cap up until it is sufficiently cool. The greenish color suggests a good shape where the coolant is. Be careful of stains and dirty-whitish color on the radiator. Once again, allow the remainder to be inspected by the mechanic such as the batteries.

Examining Under the Vehicle.

Lie down if you need to and utilize an emergency light to look at the engine beneath. Feel any indications of residue. Examine the pipes and take a look at any chance of heavy rusting.

Take it for a Test Drive.

The dealer or owner must not stop a purchaser from going over around twenty minutes of a test drive. This is a special chance to completely try to find any issues with the air-conditioning, heating unit, brakes, steering wheel, transmission, and most notably, the coziness. Feel everything; it's all right to try it on a hump or a somewhat rough roadway to truly experience the functionality. Listen thoroughly to check for any rattles. Have a paper and paper, gloves, flashlight, towel, magnet, blanket and even tape or CD when inspecting a vehicle.

Chapter 5: Before You Sign an Agreement

Security from any scams and deceitful activities needs to be a serious thing for anybody who is into lease, sales, jobs, or any sort of services which involve specific provisionary agreement and settlement.

Whether a party is a first time or a professional individual, the market and the services which go with it present the worried individual with threats and difficulties which you must not ignore. Each time that an individual is engaged in a commercial deal in the society, he/she remains in danger of submitting their selves to the likelihood of deceitful actions.

That is why agreements exist to guarantee the security of both parties included.

Typically, agreements are developed to offer strong information about the agreement which took place in between 2 or more parties and that any information mentioned therein are bound by specific laws and guidelines. Thus, it is exceptionally crucial for an individual to understand the crucial details of the agreement before he/she signs and submits to the pact.

Here is a list of a few of the things individuals ought to understand before they sign an agreement. Understanding these is going to definitely secure them from any inconsistencies or any deceitful activities which might occur.

1. Know that a contract is a legal file, bound by legal arrangements and terms.

As specified, it is a "legally binding," printed arrangement signed by 2 or more groups or factions, that involves their dedication to one another.

With the word legal, this implies that any provisions mentioned therein are bounded by law, wherein, any act, made by a specific party or all of the involved parties, that makes up as non-conformity to the details of the contract is going to be held responsible with the law.

This indicates that anyone might be locked up or held in custody unless the concerned party is proven innocent.

For this reason, it is essential to be precise about the details of the agreement prior to signing it to avoid any possible danger.

2. Make certain that you are working with reliable and dependable businesses

It is exceptionally crucial to understand initially who the concerned individual is dealing with. For that reason, it could be better if an individual or a party would attempt to do certain background checks and investigations initially prior to signing the agreement.

In case an individual or a party remains in doubt, it is ideal to go after their instincts and ignore the finalizing of the agreement.

3. Know the fine print

The issue with many folks is that they presume each detail as part and partial of the entire contract, believing that all the things are going to be specified as agreed vocally and that there isn't goint to be any damage if they do not go through the fine print completely.

This should not hold true; or else, they could get involved in trouble with the parties involved or with the authorities.

4. Make certain that all of the particulars about the agreement are completely specified.

This indicates that all information relevant to the agreement needs to be specified plainly and totally. For instance, for service agreements, make certain that the start date and ending date of the service are plainly specified.

5. Ensure that there are no blank areas in the agreement.

Prior to signing an agreement, concerned parties ought to attempt to inspect the file carefully and make sure that there are no blank areas there. This might present greater risks if left ignored, specifically if their signatures are already affixed on the agreement.

All of these things come down to the reality that individuals ought to be extremely mindful about deals, transactions, or contracts that they dedicate to. It is reasonably crucial to be conscious and well-informed of the details of the agreement prior to signing it.

You might have heard that a lack of knowledge of the law is no excuse. This holds true, for that reason, it is preferable to be completely cognizant of it than to have issues with the law ultimately.

Chapter 6: Dealership Scams

Are you considering purchasing a car but reluctant to approach your neighborhood car dealership due to the fact that you are uncertain about the procedure of purchasing a car as well as skeptical of feasible scams? Do you feel that your understanding of cars and getting great deals is insufficient, and you wish to discover more about this? Then this part is going to boost your knowledge of scams.

The Car Dealer and His Trade

To have an appropriate knowledge scams, you should initially have an idea of the car dealers themselves and how they set about their trade selling vehicles. Initially, the car dealer is going to attempt to avoid offering you a price quote. This is since the propensity of the purchaser would be to head to some other dealer and the other one is going to most likely offer a better price, therefore, the initial dealership loses the purchaser.

To boost their odds of getting you to purchase their car, they are going to attempt to make you dedicate to them prior to offeringyou the last selling price. On your end, you could do 2 things: maneuver the dealer into giving you the price quote or negotiate the deal terms. In case you chose to negotiate with the dealer then constantly watch out for these techniques:

Low Balling

This is among the tricks most frequently used by dealers. They are going to attempt to persuade the purchaser that they can offer the lowest feasible price for the vehicle, therefore, prompting the shopper to begin negotiating. The salesperson is going to most likely tell you that you can get the car which you desire for a lower amount; however, as you are ready to go into negotiations, it appears that you were not promised a better price whatsoever. It is simply a made-up promise.

Ideal Price Matching

When you say to the dealer that you are going to attempt to look around and look at other car dealerships, the salesperson is going to ask you to get back when you get the best achievable price and he is going to offer to match it. By now, you are going to most likely be too exhausted and tired of looking around. The odds are that you are going to accept their offer.

Trade-in Scam

When you have actually already purchased the car, the salesperson is going to attempt to call you before your brand-new vehicle is delivered and inform you that the first pricing of the vehicle was $500 lower and would want to gather the balance from you. In case you go for this, he is going to get an additional $500 at your cost.

Spraying

What the dealer is going to do is pursuing you non-stop up until you throw in the towel and purchase the car. Alternatively, if you chose to purchase from another dealership, he is going to call you and inform you that he might have offered you a lower price hence making you upset. To prevent this, you must not give out your contact number. Some individuals even offer incorrect contact numbers simply to dodge really annoying car dealers.

Puppy Dog Trick

The dealer is going to enable you to take the vehicle for an overnight or for an entire day; his goal is for you to fall for the vehicle and purchase it. This goes in line with the "Yoyo" scam in the following part.

Immediate Sale

The dealer is going to provide you a really appealing price; however, he is going to likewise make you think that this deal could just be made up until the day is through. This is going to make you think excessively and, ultimately, make the mistake of purchasing the car without looking at what other car dealerships can provide.

Additional Accessories

The dealer is going to offer you extras such as window tinting, car mats, and other add-ons. This intends to stop you from requesting a lesser price due to the "extras" which you are going to be getting from the dealership. It is going to slow down the price negotiations and draw away your attention far from the price.

Everybody wants the ideal offer out there, and we hate the idea that somebody would take advantage of us via modus operandi such as car scams. So prior to heading to the neighborhood car dealership, you better have a good grasp of what he is going to be speaking of. As they say, just a fool is going to go to battle unready and without any weapons. Your weapon is going to be knowledge, constantly keep that in mind.

Chapter 7: Car Financing Scams

Car dealers are frequently depicted as predators simply awaiting an unsuspecting buyer to come by. This is due to the fact that many individuals think that they are constantly on the prowl for unsuspecting purchasers that are not extremely well-informed about cars. This could be unfair due to the fact that we could argue that there are car dealers who would not lie simply to receive an added profit.

How Do You Tell the Difference?

To stay clear of ending up being a victim of sly car dealers, take a look at these car scams.

Yoyo Scam

You are going to be allowed by the dealership to bring the vehicle home as soon as you can. The dealership is going to take care of the funding, a couple of days, later on, he is going to call you once again and inform you that there was an issue with your funding plan. He is going to inform you to establish a brand-new funding plan through him, which, obviously, is going to be at a higher cost and this is going to likewise entail a really high profit for the dealer.

Watch out for this technique and stay away from it at all costs if you spot it. In case your credit standing is not good, do not have your funding done by the dealer and make plans for your own funding. If you ever do avail of the dealer's funding, you must never ever drive the vehicle back to where you live right away. Wait for at least a day simply to make certain that the processing of your funding plan has actually been finalized. By enabling one entire day to go by, you are guaranteed that the dealer is not able to utilize this scam.

Window Etching Technique

Window etching is a really frequent scam. What the dealer is going to do is to offer to engrave the VIN number of your vehicle onto the window of the vehicle for a cost. Generally, the price goes from as little as $400 to as high as $1,200. Certain buyers believe that they did an excellent job by having the ability to talk down the price to a couple of hundred dollars, however, sadly for them, a couple of hundred dollars is still a nice amount of cash. The ideal method to stay away from this type of scam is for you to purchase an etching set which you can do by yourself. This is offered in the majority of car stores and costs approximately $20. See just how much they can earn from you!

Preparation Fees

For preparing your vehicle, the dealership is going to typically include an extra preparation fee to your expense. Simply carrying out a test drive, changing fuses, or taking the vehicle's plastic cover off could have your expense going up by $500, at least! If you check out other stores, you could get the information that these add-on expenses are currently included in the MSRP as established by the producer. Certain dealerships immediately add it to the purchaser's order to make it appear necessary. When it comes to this scam, you can ask the dealer to categorize it as a credit (it ought to be same as the preparation fee amount) on the next line. In case the dealer does not accept this, you could just basically leave the car dealership.

Market Adjustment

The dealer is going to persuade you that the car you desire is selling like hotcakes and incredibly prominent. So as to sell you the car, they are going to do certain "market adjustments" totaling up to a couple of thousand dollars. This is generally shown by a tag near the MSRP tag established by the producer. Even if the vehicle you desire is incredibly prominent and is sought-after, if it is in stock, you must not be lured due to the fact that obtaining a "popular" vehicle is not worth it if you need to pay a couple of thousand dollars extra. You must never ever pay more than the MSRP established by the makers. In case you do, then you are enabling others to use you.

Warranty Extension

Even though this kind of scam is ancient, it is still being utilized and there are numerous folks who succumb to this technique. What occurs in this type of scam is that as you take a loan for the vehicle, the dealer is going to inform you that you have to buy an extended warranty due to the fact that it is among the bank conditions. There is an easy method of staying away from this scam. Ask the dealer to define plainly in writing that the extended warranty is demanded for the loan to be authorized. The dealer is most likely going to find a way to have it left out. If he continues adding the extended warranty, do not conduct business with this person and find other car dealerships.

These are a few of the most typical car funding scams that are used by some car dealerships. Constantly keep these in mind if you will purchase a car. If you or a buddy were treated reasonably by a dealer before, think about utilizing the identical dealership once again. It's an excellent sign that they do appreciate their clients and aren't simply looking for a "quick buck."

Think really cautiously and do not purchase impulsively.

Chapter 8: Test Driving Suggestions

Are you looking for tips on what to look at when thinking about purchasing a car? It is vital to have a comprehensive understanding of the car that you want to get. This is going to guarantee your pleasure and satisfaction for many years ahead.

The Appropriate State of Mind

First off, you need to remember that test driving a car is not an easy procedure. It is going to identify the declared functionality of the vehicle you desire by the dealership and compare it with the functionality in actual conditions. This is the time in which you can see if the vehicle which you have always desired matches up with your needs and ideals.

Have a Checklist of Requirements

It is essential to create a list of requirements by which you are going to have the ability to evaluate a vehicle's functionality without having any issues concerning objectivity. It is going to aid you to see the car's benefits and drawbacks without the impact of other elements which are not as essential. This is going to additionally allow you to carry out the test drive much faster, given that you currently have a set list of the things which you are trying to find in a vehicle.

Explore Other Weather Conditions

You ought to think about test driving a car in harsh weather to have a much better feel of the vehicle's general functionality despite the kind of weather it is going to be utilized in. It is great to understand the car's feel when driving in rain or during the night. If the dealer demands that a representative is with you during the test drive, allow the individual understand what you intend to do and what path you are going to be taking.

Drive in Various Sorts of Terrain

During a test drive, lay out a path that is going to take you to various types of terrain such as steep hills and rough roads. When inspecting a vehicle's braking power and turning capabilities, think about doing it on a side street which has extremely little traffic. In case you intend to assess the vehicle's acceleration and velocity, then taking it to the interstate or the open highway or is a great idea. You need to additionally do a great deal of driving on the kinds of roads that you come across in daily driving like your path to work.

Car Interior

As you are behind the wheel and within the car, a few of the things which you must check are these:

Try to see if the gauges are easy to read and working correctly. Make sure you are able to quickly see beyond the steering wheel and that it does not block your view in any way.

Dashboard

After steering the wheel and inspecting the gauges, the following thing to take a look at is the dashboard. See that all the dashboard controls are within grasp and that you do not have to move out of a comfy driving position substantially. Look for any dashboard extensions which might possibly lead to injury to the passenger or driver throughout unexpected stops or when surging forward.

Visibility

Inspect that windscreen wipers are functioning correctly. Utilize the spraying system that administers the wiper liquid onto the windscreen and discover areas which might not be reached by it. This is crucial considering that it might impact your bad weather and night time driving. Is there a tint? In case the tint is too dense, it is going to hinder night visibility and induce you to have a tough time driving during night. There are likewise laws which stop the windscreen from being tinted. Examine the guidelines and policies in your state to figure out if that applies.

On the Road

While driving, look for blind spots by carefully taking a look at the side mirrors in addition to the rearview mirror. Assess the functionality of the vehicle's suspension by inspecting if it provides a comfy ride and if you feel that you remain in control constantly. Try to find a parking area without too many cars and perform a couple of sharp turns along with a U-turn. Does the suspension function properly or does the car wobble excessively?

Steering

Check if the vehicle has a responsive steering system. A small turn of the steering wheel ought to suffice for the vehicle to react effectively. The steering ought to be sufficiently balanced in order to have ample power for easy steering, and simultaneously, it ought to properly convey the road feel to the driver.

These are a few of the things to think about when taking a vehicle for a road test. Keep in mind to place a car's functionality as a priority ahead of other elements.

Chapter 9: Purchasing Cars Online

The internet is a hassle-free location to purchase cars. There are lots of benefits to shopping for cars on the web. Initially, it significantly decreases the inconvenience of dealing with salespersons and car dealers. When purchasing a car on the web, you do not need to listen to a dealer talk about the specifications of a car. You simply need to read all about it on a site. You do not need to go to a display room, you simply need to use your mouse and you can browse through various makes and car models. And if you wish to take another glance at a car, it's really simple.

The benefit of shopping for a vehicle on the web is that all deals could be done without you needing to go out of your home. After buying the car you desire and paying the price, all you need to do is to wait for your brand new car to be shipped to the closest dealership. Certain dealers would even ship the car straight to your doorstep.

There are essentially 2 kinds of online car shopping sites. One is going to just get you in touch with an actual car seller. This suggests you need to transact the conventional manner. Then there are sites that are going to handle everything. Some sites would even ship the vehicle to you. Naturally, this would involve a larger amount of service fee. Shipments are made just to close-by locations.

However, before you click and purchase, you need to initially understand precisely what sort of car you desire. Figure out the kinds of cars you're searching for and the price you want to pay.

Beneath are certain other suggestions on how to purchase a car on the internet.

1. The most crucial thing to do is to research. Do not get too thrilled. Check out the various plans out there.

2. Make certain that the site you go to is safe and dependable. The majority of websites would allow you search by price range, car type or both.

3. If you do not wish to make all deals on the web, then you might ask the dealer to meet you face to face as soon as you discovered him on the web. Upon meeting the dealer, negotiate with him as you would with a regular car dealer and after that, sign the documents.

4. It's not required to pin the year, model and make of the automobile you wish to purchase. Simply a basic photo would do. Figure out how you are going to be utilizing your brand-new car and why you are purchasing a vehicle to begin with. Ask yourself what qualities are essential to you.

Do you provide more significance to fuel efficiency instead of speed or vice versa? Are you more worried about security features than an excellent stereo? After you have actually done all these, it's time to identify your budget plan. Simply just how much are you going to spend on a car?

5. As soon as you have actually chosen a brand name and model car, it would be wise to look at the history of the specific car maker. You could quickly do this by looking at sites and consumer magazines.

You might likewise have a look at the site of the cart's producer (e.g., Toyota, Ford, BMW, and so on).

6. If you can not pay for brand-new models, recall a few of our prior discussions. There are sites which offer used cars. Prominent search engines typically have an automobile section, so this is an excellent location to shop. If you do not find one here, you could attempt looking by utilizing the search term "used car."

7. Inspect the background of the site of the internet car dealership by clicking on the "about us" area of the site. It is recommended to purchase just from sites that have comprehensive information about the cars they are selling. Ask if they have actually carried out evaluations of each vehicle in their lot via independent mechanics. In case they have, then you'll understand that site is a great location to buy cars. A lot of producers' websites have actually comprehensive information on models, consisting of offered choices, images and MSRP (Manufacturer's Suggested Retail Price).

8. Check out reviews about various models, makes and years. An excellent site should likewise have reviews and scores for the vehicles they are offering. However, you need to additionally cross-check by checking out independent vehicle websites or websites that do not offer cars or are linked to car businesses to get more unbiased reviews.

Chapter 10: Do not Purchase a Flood-damaged Car

Do not purchase flood-damaged cars. There are numerous flood-damaged cars offered on the marketplace. Typically, flood-damaged vehicles, trucks and SUVs are put in salvage yards.

Some, however, are created spick and span and placed into the car market and they are going to be up for sale. However, the issue is, there actually are no indications of flood damage which would give them away. This is since the car is going to be restored so they would appear practically new.

The damage brought on by the flood would be covered or removed. After the modifications, the cars are then going to be offered to unsuspecting shoppers who believe they are obtaining an excellent deal.

Reality is that flood-damaged vehicles are being moved by deceitful merchants. Consumers can and ought to secure themselves from being taken advantage of in the car market. Fortunately, there are a number of things people could do to shield themselves from purchasing flood-damaged vehicles.

The ideal thing a customer could do to be certain if a vehicle is flood-damaged is to get the history of the car. You could do this by sending the identification number of a vehicle to a site that supplies car history information. What these sites do is search a nationwide car information database to gather research on the car's title, upkeep record, odometer and registration. You are going to instantly understand if the car has actually been stolen, has actually had a troubled past, or has actually had its odometer rolled back because of the report that the site is going to come back. Simply a bit of research on a car's history is going to show if it has actually been restored, flooded or rebuilt.

Potential car purchasers ought to be aware of flood-damaged cars being offered on the market. Why shouldn't you purchase flood-damaged vehicles? Well, simply since water leaves long-term damage. Even if the machines and devices needing electrical energy are going to restore it, it is going to most likely stop working eventually since mildew and mold aren't' simple to get rid of.

A few of these suggestions are covered somewhere else; however, they can't be stressed enough, particularly regarding a flood-damaged car. Beneath are numerous things you may do to examine if a vehicle is flood-damaged:

Look for Dirt and Moisture

Flood-damaged cars generally have dirt inside the lights and trapped moisture. Moisture can likewise be seen within the gloves compartment, trunk and console, so you better examine these areas. Dirt, which can likewise suggest flood damage, can likewise build up beneath the hood. Wetness can likewise build up beneath the seat. Naturally, rust is another indication of flood damage.

Smell the Car

Mildew could be quickly discovered by smell. Mildew frequently forms on drenched fabrics, so hone your smell sense as you're on the lookout for a brand-new car. Likewise, try to discover other smells which may be brought on by flood damage such as spilled fuel or oil.

Inspect if Parts Match

A mismatched part might suggest that the parts are changed fast after the vehicle has actually been salvaged from a flood. So attempt to see if the seats, carpet and stereo parts appear too fresh for the vehicle.

Likewise, try to inspect if the car has actually been titled numerous times from various states, which is generally a clue that its owners are attempting to remove the questionable and unfavorable history of the vehicle by trying to find areas where disclosing flaws is not needed or is quickly evaded. Cars which are titled a number of times are typically totaled or salvaged.

Test Drive

Obviously, the ideal way to examine the functionality of a vehicle is to take it for a test drive. Examine the electrical system consisting of the lights and the stereo.

Ask a Professional

Have a professional technician or mechanic examine the car. Have another opinion if you can. Professional car technicians and mechanics could discover flood-damaged vehicles quicker than regular people.

Keep in mind that when purchasing a car, you should never ever gamble. Purchasing a damaged car could cost you more than your cash. It might likewise bring a major accident, even death. If you think that somebody is offering you a car which has actually been flood-damaged, instantly say no and then leave. The cash you are going to save in purchasing a flood-damaged vehicle is going to rapidly replace the headache it is going to bring.

Chapter 11: Things to Be Careful About When Purchasing Second-Hand Cars

These days, life could be quite hard regarding cash. Everything is going up, with prices rising to absurd heights. In addition to high expenses, it's no surprise why there are more scams, rip-offs and deceptive activities. Individuals would like to know they can get an excellent "deal" which makes them susceptible to those who are seeking to make a quick buck!

With all these tricksters and fraudsters lurking about, it is ideal to constantly be on guard and understand the things to be prevented so as to protect against the likelihood of being a victim of swindles and scams.

Subsequently, individuals who are purchasing used cars need to be aware of the various deceitful activities being utilized by dishonest individuals so as to have the best deal with used cars.

Additionally, when it comes to individuals who know that the car they purchased is stolen, odds are, they are going to be apprehended and are going to be lawfully held responsible.

Here are certain things to stay away from when purchasing used cars:

1. Used car purchasers ought to stay away from any deals which are "too good to be true." This is going to just make things worse if the shopper is going to think that the offer is the best-used car offer of all time.

2. Shoppers ought to stay away from purchasing used cars from sellers who do not offer a permanent address or the actual workplace of the phone number provided.

3. It is essential for a customer to inspect the VIN or the car identification number plate. It should be firmly attached to the utilized car's dashboard, without any rivets which are loosened.

Loosened up rivets would suggest that the VIN plate does not fit or it has actually been formerly gotten rid of.

4. Likewise, the buyer must stay away from purchasing used vehicles which have VIN plates which are spruced up, paint is freshly redone, and the numbers appear as if they are not the authentic "factory numbers."

VIN plates can be quickly swapped by a crook by using those that are taken from a ruined car.

5. As much as feasible, it would be ideal to stay away from purchasing a used car which is newly painted. There are circumstances in which the stolen car's identity is being altered by changing the color of the paint.

6. Used car shoppers must stay away from purchasing cars from a seller who can not present the vehicle's "insurance policy." This might indicate that the car is stolen or that the person selling is not the actual owner of the vehicle.

It is incredibly crucial for the purchasers to take note of these things prior to purchasing a used car. As they claim, prevention is the ideal defense.

Chapter 12: Auctions

Car auctions are a great location to get great bargains on vehicles. However, purchasing cars from auctions could be really convoluted. This is especially true for the inexperienced. Naturally, cars in auctions are usually cheaper. However, they are cheap since they do not have the guarantees which come with cars purchased from dealerships. The auction rules are straightforward, you bid a product, you pay, and it's yours.

It is, for that reason, not recommended for individuals without a lot of experience in motor vehicles to simply go to an auction and purchase a car. What you'll require if you intend to purchase a car in an auction is a good eye, excellent observation abilities and a desire to learn the car auction procedure.

You could even make it a business by purchasing vehicles at auction, and after that, offering them at a profit. There are a number of individuals who have actually made substantial profits by transacting with vehicles at auction.

There are essentially 2 sorts of car auctions, online and local. Each of the auctions function in a different way. Each has various sets of guidelines, so you should study them really thoroughly if you intend to purchase a car from either sort.

If you wish to purchase a car in a car auction, you need to initially pre-register with the local auction you wish to sign up with. The pre-registration procedure will not take a lot of time and is going to offer you all the vital information that you are going to require so as to purchase a car. An auction website is going to frequently have auction personnel who are going to describe to you the guidelines of their auction website and the procedures and direction on how to take part and bid on the vehicles.

It is smart to check the cars prior to the actual bidding starts. Ask the auction organizer or personnel if they can enable you to take a more detailed peek at the cars prior to the beginning of the auction itself. Another smart decision is to establish your payment limit prior to the sale. By doing this, you will not succumb to the overbidding trap.

As in other auctions, it is extremely crucial that you signal your quotes cautiously to the auctioneer to stay away from confusion. Settle your deal as quickly as the auction concludes. Search for the block clerk to settle your accounts. A car auction staff are going to generally be there to help you when you are prepared to pay for the car you have actually won and should likewise have the ability to ensure you a great title for the car.

You need to likewise inspect the car after you have actually won. Take the car for a test drive and ensure that it matches the representation and guarantee made throughout the auction.

A lot of auction websites would take credit card or check if you can not pay the vehicle with money. The auction staff are going to additionally generally provide you a third party funding offer. It is a good idea though to have your own finances prepared prior to going into a car auction. Outside funding is generally more affordable than the majority of in-house funding of auction sites.

Request an invoice copy from the auction staff. This is going to allow you to drive or deliver your freshly purchased car to your place. A lot of vehicle auction supervisors will offer you good deals on shipping arrangements, so ask about this.

There actually are no substantial distinctions between a neighborhood auction and an internet auction. Car auctions on the internet function basically identically as local ones. The only drawback of internet auction websites is that you will not have the ability to check the vehicle as you would in an actual auction. In taking part in internet car auctions, ensure that you are going to be offered with all the required details about the car upfront and that there is an actual photo of the car. Vehicles purchased from an online auction are normally delivered to the purchaser for a minimal cost. Check the car right away upon arrival at your area to reduce risks.

Chapter 13: Fuel Efficiency Suggestions

Boosts in the gas price, accompany the boost in the amount of fuel-saving frauds.

A constant boost in the gas price is frequent news these days. This is accompanied by a rise in advertising exposures for "gas-saving" devices. These items bring in prospective purchasers looking for approaches to optimize fuel efficiency.

There are necessary processes car owners can take into consideration to enhance gas mileage. Based upon the BBB, the general public ought to be critical in examining products which declare gas-saving features for car gadgets or additives for oil and gas.

There are items which feature gas-saving attributes that truly work. Nevertheless, customers could be dealing with a significant engine problem or a nullified creator's warranty by placing devices to the engine.

Customers need to watch out for these particular advertising mottos: "20 percent fuel efficiency enhancement."

A hundred or more devices which include gas-saving properties are shown untrue by the Environmental Protection Agency. Worse, there are so-called "gas-saving" gadgets which might cause a damaging impact on a car's engine.

One more side effect are additional smoke emissions. Instances of these devices are Additives, Engine Modifiers, Liquid Injection, Fuel Line Gadgets, Vapor and Air Release Devices and a great deal more.

" Get an extra 4 miles for every gallon with this item."

Customer reviews are prevalent in fuel-saving ads. Presuming these testimonials are not comprised, a couple of customers evaluate the fuel usage of their car prior to putting on the device. For that reason, an unbiased contrast of the fuel usage prior to and after the product was included is not going to be achieved.

As the alternative in searching for gas-saving devices, the BBB recommends that the general public must take into account doing more than a single activity which could aid save gas. The crucial way to start is to focus on the gas pump. The customer ought to purchase just gas as required. Make certain to examine the vehicle handbook to understand the appropriate level of the car octane.

The following are useful suggestions to save gas:

Simply drive reasonably within the speed limitation. Keep in mind that speeds of more than 60 miles per hour boost fuel usage.

Do not do unexpected starts, stops and accelerations. Acceleration must be done slowly. The gas pedal must not be stepped on more than one-fourth of the way down. This is going to enable the engine to operate more efficiently. Gas conservation might rise to 5 percent if abrupt stops, accelerations and jerks are avoided.

Utilize gears for cruise control and overdrive as required. Fuel efficiency is obtained when being on the highway.

Windows ought to be closed on highways. Open windows could result in air drag which could reduce fuel usage by 10%.

Rough roadways must not be taken as much as feasible. Bumps, dirt, gravels and rough roads can result in a 30% boost in fuel intake.

Take out unneeded luggage. The trunk must be kept tidy, and gizmos, tools or loads which are not required ought to be taken out. 100 pounds of additional luggage can impact fuel economy by 2 percent.

The vehicle must always be maintained. The engine needs to be given a routine tune-up, the tires ought to constantly have the appropriate air pressure and be effectively lined up, the oil ought to be changed when required and air filters should be changed frequently. Blocked filters could impact gas intake negatively by as much as ten percent.

Shut off the engine whenever there is a chance. Idle engine squanders fuel. There are circumstances where the engine may be shut off; waiting for somebody, stopping for gas, modifying tire pressure, being stuck in traffic and a great deal more.

Engine warm-up fuel conservation suggestions:

The prolonged heating up of the engine must be avoided. 30 to 45 seconds ought to be sufficient time.

Inspect if the automatic check is eliminated after heating up the engine. This is generally stuck, which might typically result in bad air and gas mix.

Do not rev the engine. This is typically done prior to switching off the engine. This results in unneeded loss of fuel and likewise washes the oil inside the cylinder walls. Consequently, there is air pressure loss and subsequently fuel loss too.

Chapter 14: Read the Warranty; It is Going to Conserve Some Cash

When consumers purchase a vehicle, the seller or maker promises to vouch for the car. This is a warranty. Federal law imposes warranties to be constantly available to shoppers. It must be made apparent to the customer even when they are simply looking for which car to purchase or if they are simply surfing the web.

Coverage varies. For that reason, warranties need to be examined along with the value, design, quality and other car qualities.

Guarantees and warranties frequently puzzle individuals that are attempting to distinguish the two. These 2 words, although rather comparable, have unique differences.

Initially, the fundamental terms are rather different. Warranty is generally credited to producers while guarantees describe the labor.

Warranties are generally pro-rated. This indicates that the coverage agreements might alter as time goes on. Guarantees from the labor sector mainly are inclusive.

Here are some instances:

1. Siding must have a warranty not to break, buckle or sag.

2. The strength of colors is not going to lessen for 3 points or more.

3. Warranties are transferable to various owners, and after that, altered to 5 years.

Impacts of stains produced by chemicals or excessive mildew as a result of the fact that the requirement for cleaning is not a part of the warranty.

Warranties typically provide satisfactory coverage but are not perfect.

What is the Coverage of Guarantee Under One Year of Labor?

A service or product trips up in just the initial year. It is going to be replaced or repaired without charge. In short, if a high priced item is bought, the coverage is going to be 100%. As item cost declines, so does the coverage value. It is not the producer's fault that less coverage is offered for low-cost goods.

Product information and guarantee or warranty additions must be completely comprehended by the prospective purchasers prior to signing a thing. Constantly watch for professionals that stick to particular brand names for a very long time. Odds are they have legitimate reasons why they stay faithful.

Everybody is like a loyal contractor who tries to find excellent service. These contractors are constantly on stand by to provide service. Warranties and item quality worries likewise keep great contractors on their toes. However, bad contractors do not care whatsoever for warranty and product quality. They are constantly searching for the least expensive product on sale and completely ignoring the services and warrantees which it might entail in the future.

What Are Written Warranties?

Warranties in writing are not compulsory under the law yet are typically discovered in significant purchases. Here are certain suggestions to think about in trying to find warranties.

What is the Amount of Time Offered By the Warranty?

Constantly ensure to take note of the start and expiration of warranties and, additionally, the terms which might nullify it.

Who is The Person in Charge of Supplying the Warranty?

The coverage is going to be offered either by the producer or seller.

What Are the Particular Actions to Be Carried Out When There is a Product Failure?

Inspect if the company, seller or provider is going to repair the item, change it, or provide the cashback.

What Are the Parts and Particular Product Failings Covered by the Warranty?

Purchasers ought to completely investigate if there are product components or kinds which are not an aspect of the warranty coverage. Instances are warranties which request payment for work done throughout the repair procedure. Watch out for additions that could possibly be pricey or troublesome.

Are "Consequential Damages" Covered by the Warranty?

There are warranties which do not consist of damages impacted by the product or the time and cash invested in fixing the damages. An instance of this is when a freshly purchased freezer breaks down and the food is ruined. The seller or maker is not obliged to pay for the food.

What Are the Warranty Limitations?

There are warranties which just provide the warranty coverage if the product status is preserved as mentioned in the condition. Specific warranties define products such as washing machines to be utilized for home usage only. If the washing machine is utilized in a company and breaks down, no warranty is going to be provided.

Buyers ought to constantly go through the guarantee or warranty terms prior to buying any product in case they wish to maximize their hard-earned cash.

I hope that you enjoyed reading through this book and that you have found it useful. If you want to share your thoughts on this book, you can do so by leaving a review on the Amazon page. Have a great rest of the day.

Made in the USA
Columbia, SC
29 August 2024

41261911R00048